Examining
Water Power

Rebecca Cooke

CLARA
HOUSE
BOOKS

First published in 2013 by Clara House Books, an imprint of The Oliver Press, Inc.

Clara House Books
5707 West 36th Street
Minneapolis, MN 55416
USA

Produced by Red Line Editorial

The publisher would like to thank Michele Guala, assistant professor in the Department of Civil Engineering and St. Anthony Falls Laboratory at the University of Minnesota, for serving as a content consultant for this book.

Picture Credits
Belinda Pretorius/Shutterstock Images, cover, 1; Martin Lehmann/Shutterstock Images, 5; Shutterstock Images, 8, 25, 26–27, 41; Georgios Kollidas/Shutterstock Images, 9; Library of Congress, 11; Jeremy Mayes/iStockphoto, 13; Matej Hudovernik/Shutterstock Images, 15; Richard Gillard/iStockphoto, 16; Terry Davis/Shutterstock Images, 17; Dani Vincek/Shutterstock Images, 19; Todd Paris/Alaska Power & Telephone Company/AP Images, 20; Red Line Editorial, 21; Bryan Brazil/Shutterstock Images, 23; Robert F. Bukaty/AP Images, 28; EpicStockMedia/Shutterstock Images, 31; HO/Finavera Renewables/AP Images, 32; Warren Goldswain/Shutterstock Images, 34; Galyna Andrushko/Shutterstock Images, 36; National Oceanic and Atmospheric Administration/Department of Commerce, 39; Jaren Jai Wicklund/Shutterstock Images, 45

Library of Congress Cataloging-in-Publication Data
Cooke, Rebecca, 1978-
 Examining water power / Rebecca Cooke.
    pages cm. -- (Examining energy)
 Audience: Grades 7 to 8.
 Includes bibliographical references and index.
  ISBN 978-1-934545-46-1 (alk. paper)
 1. Water-power--Juvenile literature. I. Title.

  TC146.C66 2013
  333.9'14--dc23

                          2012035317

Printed in the United States of America
CGI012013

www.oliverpress.com

Contents

CHAPTER 1

The Power of Water

Have you ever heard someone in your family complain about the high cost of electricity? People around the world use a lot of energy, and the demand for energy continues to rise. Right now, most of this energy comes from non-renewable sources, such as fossil fuels. Fossil fuels, such as oil, coal, and natural gas, can be bad for the environment. They also take millions of years to form, which means we are using them faster than they can be produced. Because of this, many scientists and innovators are trying to find ways to increase our use of alternative energy sources. Alternative energy sources include biofuels, wind energy, solar energy, and water power.

Water covers more than 70 percent of the earth's surface. About 96.5 percent of Earth's water comes from oceans. Only 2.5 percent of Earth's water is freshwater. It comes from

Moving water is a renewable resource we can use to generate energy without burning fossil fuels.

groundwater, glaciers, ice caps, lakes, and ice and snow melt. A tiny fraction is water vapor in the atmosphere.

We can harness the energy from moving water, such as rivers, ocean waves, and tides, to create energy we can use. Energy from moving water is clean and renewable. Water is not consumed during energy generation the way fossil fuels are when burned to create energy. Water can be used over and over again.

Converting water power to electricity does not pollute the environment. However, building hydroelectric facilities and manufacturing hydroelectric equipment, such as turbines and dams, does produce pollution. Still, the availability of water means it will play an important role in our future energy production.

EXPLORING WATER POWER

In this book, it is your job to learn about water power and its place in our future energy production. How did people use water power in the past? What kinds of moving water can we use to create energy? What kind of energy does water power give us? How do we use it now? How might we use water power in the future?

Zach Miller is part of a study group researching alternative energy for a science fair project. Zach's job is to learn about water power. He is conducting research and interviewing experts in the field to learn more about water power as an energy source. Reading Zach's journal will help you in your own research.

2,500 Years of Water Power

I decide to start my research by discussing how past cultures have used water power as an energy source. I stay after school to talk to my science teacher, Ms. Arnold. I'm surprised by how much she knows about the history of water power.

"Water power may be considered an alternative energy source today," Ms. Arnold says, "but humans have been harnessing the power of water for longer than you might think! Ancient records are fuzzy. But historians believe the ancient Chinese used water power to grind grain as early as the 100s BCE. Romans were the first to write about water-powered wheels, although it seems as if they didn't use them often—at least not until the 300s CE. Then they used 16 waterwheels to grind grain in what is now southern France. Waterwheels were an important source of energy throughout Europe from the 1000s on."

Humans have harnessed the energy of moving water for more than 1,000 years by using waterwheels such as this one in the Czech Republic.

Ms. Arnold picks a small pinwheel up off her desk as she talks. "But it doesn't really matter who used water power first. What's important to know is that waterwheels operate much like this pinwheel I'm holding."

She blows on the pinwheel, and it spins.

"Air turns this pinwheel much like water turns a waterwheel. The mill uses the kinetic energy, or energy of motion, of the flowing water to turn the stones that grind the grain."

Ms. Arnold tells me that there are two basic types of waterwheels. They are called undershot and overshot. The undershot waterwheel is partially submerged in the river. Its grinding speed depends on the speed of the river as water flows by. A slow flow means slow grinding.

With overshot wheels, the water source comes from above. Gates above the wheel, known as sluice gates, control the flow of the water, which turns the wheel. The wheel drives a complicated set of gears that perform work, such as grinding wheat into flour, powering a sawmill, or pumping water.

"Thanks to Michael Faraday, a British chemist and physicist, hydropower, or power harnessed from water power, became a force for generating electricity in the late 1800s," Ms. Arnold tells me. "His experiments led to the first generator to produce a steady electric

Michael Faraday discovered that when a magnet was moved near a wire, it created an electric current. This discovery led to the technology that then led to hydroelectric power plants.

current. His findings became the basis of hydroelectricity. In 1882, the world's first hydroelectric plant opened in Appleton, Wisconsin. By 1940, hydropower had become super popular. At that time, it provided 33 percent of U.S. electricity."

Ms. Arnold explains that in the late 1940s, it became cheaper to produce electricity with fossil fuels than hydropower. Fossil fuels took over as our primary source of electricity generation. Then, in the 1970s, the United States experienced a shortage of petroleum for gasoline. This shortage sent us scrambling for alternative sources of energy, such as water power.

Burning fossil fuels releases carbon dioxide and other gases into the atmosphere. These gases trap the sun's heat similar to how heat is trapped in a greenhouse. This is known as the greenhouse effect, and the gases are known as greenhouse gases. Many scientists

THE TENNESSEE VALLEY AUTHORITY

In the 1930s, the United States went through an economic depression. Many people were out of work and living in poverty. In 1933, the U.S. Congress created the Tennessee Valley Authority (TVA) to help improve standards of living along the Tennessee River, which included areas of Tennessee, Kentucky, Alabama, Georgia, Mississippi, North Carolina, and Virginia. The TVA built dams on the Tennessee River and its tributaries. These dams helped control the frequent floods that occurred along the river. They also generated power that provided badly needed electricity to the region. The new power plants also brought many new jobs to the region. Today, the TVA provides power to more than 9 million people in an area covering more than 80,000 square miles (200,000 sq. km).

The Tennessee Valley Authority constructed the Fort Loudon Dam in Knoxville, Tennessee, from 1940 to 1943.

believe adding more greenhouse gases to the atmosphere is increasing the greenhouse effect. This, they believe, is contributing to a gradual warming of the earth's average temperature. That's why water power is such an attractive option. Fossil fuels are still an important part of our energy production, but it's important to keep considering alternative energy sources.

CHAPTER 3

At the Dam

I bike along River Parkway, a path for cyclists and walkers. To my left, the river tumbles over rocks in swift-running currents. Ms. Arnold said that fast moving water has kinetic energy, and I know just the place to harness it for electricity: the dam.

Jill O'Malley, the dam's manager, invites me to the top of the dam. It sits 340 feet (104 m) above the river. "Dams are one of the oldest ways to create electricity from moving water," Ms. O'Malley begins. "They have done a lot to provide power and control water across the United States. Dams hold back river water. At first, many dams were built to control floods and water crops. Today, many dams are used to generate hydroelectric power. The Hoover Dam on the Colorado River creates so much power that the dam can be maintained using only the money made from power sales. Lake Mead, which was created when the river was dammed, is one of the largest man-made lakes in the world. Like many lakes created by dams, Lake Mead is a reservoir for human use. The lake provides water to Arizona, Nevada, California, and even northern Mexico! The lake also provides recreation opportunities like boating."

"Can a dam be built on any river?" I ask.

'ROUND AND 'ROUND THE WATER CYCLE

Water is the only natural substance on Earth found in liquid, solid, and gas form. Water constantly shifts between these three states whether on, above, or below Earth's surface. The sun heats surface water and evaporates it. The evaporated water condenses into clouds. Precipitation, such as rain or snow, falls on the ground and mixes with our rivers. River water flows to the oceans or lakes, where it evaporates to repeat the cycle.

Ms. O'Malley shakes her head. "First, we need a river with a steady flow that receives regular precipitation in the form or rain or snow. Second, we need water from mountains and surrounding areas to drain into the river and continue feeding it. Third, we need an elevation drop, such as a waterfall. The higher the drop, the greater the energy potential of the river."

I peer over the edge of the dam to the river below. "How does the dam produce electricity?"

Ms. O'Malley smiles. "The dam itself doesn't produce electricity. The dam is a concrete structure that backs up the flow of the river and creates a lake." She points to our left. "Think of the lake as stored energy. When the dam's gates, known as sluice gates, open, water from the lake rushes down the penstock, or pipe, to spin the turbine. The turbine converts the river's kinetic energy to mechanical energy that runs a generator. The generator converts the

AN ENGINEERING MARVEL

The Hoover Dam was built between 1930 and 1936. It was named after President Herbert Hoover, who played a big role in getting the dam built. Each year, melting snow from the Rocky Mountains caused the Colorado River to flood. This flooding hurt farming towns along the river. The dam helped control the flooding. It also created a reservoir that could provide water across the Southwest. Americans could farm land they hadn't been able to before the dam. The dam also produced electricity for locations across the Southwest. At 726 feet (221 m) high, the Hoover Dam is the second-highest dam in the country. It provides water and power to millions of people. The dam is still an engineering wonder.

The Hoover Dam helped make it possible for many people to live and farm in the southwestern United States.

mechanical energy to electricity, which is then sent over power lines to your home."

I look at the water tumbling over the dam. "What happens to the water?" I ask.

Ms. O'Malley rubs her hands together. "That's the beauty of water power! Unlike biofuels, which are grown and processed, and fossil fuels, which are mined or drilled, water returns to the river on the downstream side of the dam. It's a free and renewable fuel source that never produces greenhouse gases. Hydropower simply borrows the energy in the water before returning it to the river."

I look at the lake on one side of the dam and the river on the other. "You mean this lake wasn't here before the dam?" I wonder if all of the windsurfers and water-skiers enjoying the lake today know that.

"Right. The area used to be a river with banks and rocks much like you see on the downstream side of the dam. When the dam was built, the lake flooded everything in its path."

Ms. O'Malley explains that building the dam definitely changed the natural environment of the riverbed, both upstream and downstream. It destroyed the homes of many animals living along the riverbanks. She says that dams also change the water chemistry. The oxygen level of the lake is lower than in the free-flowing river. This change affects the plants and animals living in the river. Fish, such as salmon and steelhead trout, used to migrate upstream for centuries

Dams use energy of moving water to create electric power. They also create large reservoirs people can use for fun activities, such as fishing.

Salmon jump up a fish ladder's steps one at a time. At the top of the ladder, they swim into the lake on the upstream side of the dam.

to lay their eggs. Now the dam is blocking their path. Fish must climb a special ladder over the dam to swim further upstream.

Dams also impact the environment by blocking sediment that would otherwise be transported by the river. This trapped sediment slowly fills the lake created by the dam. This blocks sunlight and makes for murkier water, changing the environment for the plants and animals.

I never thought about how water power might affect plants and animals. I guess there are pros and cons to water power, too.

CHAPTER 4

The River Wild

Today I'm about to add another piece to the puzzle. Water storage is a big deal for hydroelectric power plants. If the water is stored in the form of a lake on the upstream side of a dam, electricity can be generated at will. But what if the water isn't stored? Can electricity still be generated?

To answer these questions, I hop on a bus headed for the busy river docks. Our town has one of the few rivers still used for shipping. Joseph Trey has offered to meet me there. He is one of the founders of a company that is making electricity from the river's current. I spot Mr. Trey wearing his ball cap and a Windbreaker. We board a barge with a giant crane and something that looks like a three-bladed ceiling fan lying on its side.

"You picked a great day to stop by, Zach," Mr. Trey says. "We're deploying another turbine in the river today." He points to the odd-looking fan. "We studied this river very carefully. The size of the turbine's blades is determined by the depth and the speed of the river—the deeper the river, the larger the blades."

"Aren't they in the way of ships that travel the river?" I ask. "And what about the fish?"

"Good questions," Mr. Trey says. "We anchor the turbine to the floor of the river away from the shipping channels. Debris floating on the top of the river floats right over our turbines. Fish and other marine life might get a gentle nudge, but the turbines spin too slowly to injure them."

"How many turbines will you put in this river?" I ask. The river is too murky for me to see the bottom.

Mr. Trey grins. "That's a great question! It's actually something we are still trying to figure out. We are currently testing the effects of different numbers of turbines in the water. More turbines do not always mean more electricity. We also need to

Rivers can move extremely fast. If we can capture this energy, we can use it to generate electricity.

River turbines come in all shapes and sizes. This turbine being installed in the Yukon River in Alaska can generate electricity without using a dam.

determine the best spacing for the turbines. Too close and the turbines interfere with each other. But too far apart, and we waste some of the kinetic energy of the river."

"Mr. Trey, your turbines are really cool. I can see they don't pollute the river and don't create greenhouse gases. But how much electricity do they produce? "

"You and my investors ask the same questions," Mr. Trey says. "An underwater cable connects our turbines to the power grid, where the electricity is distributed to homes and businesses. Each turbine in this river produces about

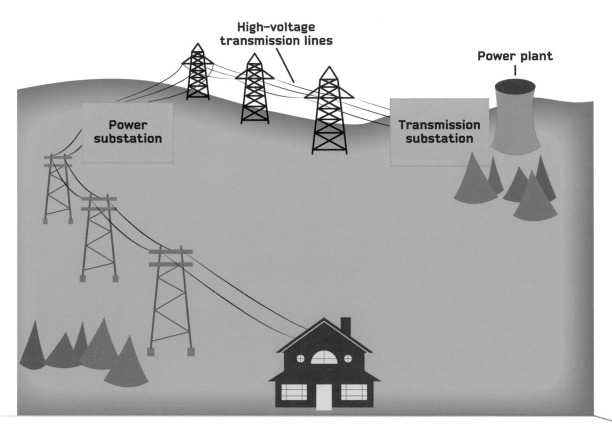

DEMYSTIFYING THE POWER GRID

The power grid transports electricity from its source to your home. A hydroelectric plant turns the energy from waves, tides, or dams into electricity. From the plant, electricity travels to a transmission substation that increases the voltage for long-distance travel. When the high-voltage electricity arrives at another substation approximately 300 miles (480 km) away, its voltage is decreased for use in your home. In the United States, people typically use appliances that require 120 to 240 volts of electricity.

RESTORE THE HETCH HETCHY VALLEY?

Between 1913 and 1934, the city of San Francisco, California, built the O'Shaughnessy Dam across the Tuolumne River in Yosemite National Park. The new dam covered the Hetch Hetchy Valley floor with 300 feet (90 m) of water. Today, San Francisco relies on this dam for its water. The Restore Hetch Hetchy movement would like to destroy the dam and return the river valley to its former state. Doing so would restore animal habitat and improve the scenery of the valley. But the city of San Francisco would lose its main source of water. What do you think? Should the dam be taken down?

35 kilowatts of electricity. In contrast, the dam you visited yesterday produces about 200 megawatts— that's more than 5,000 times as much energy. But dams require a substantial change in the environment to build and operate. Our turbines are much less intrusive."

I had no idea so much power could come from rivers. I wonder how we use other bodies of water as energy sources.

The dam that created a reservoir in the Hetch Hetchy Valley is very controversial.

CHAPTER 5

The Ebb and Flow of Tides

Today I'm visiting another river that uses tides instead of currents to generate electricity. I'm meeting with Jessica Howard. She studies tides, and I'm hoping she can give me more information about how we use tides to generate power.

Ms. Howard jumps right into the history of tidal power. "More than 1,000 years ago, people ground grain using waterwheels powered by tides instead of river currents. When the tide flowed in, these devices captured tidal water much like dams hold water. The captured water was released to turn the wheel that moved the grinding stones. Fast forward to today. Turbines have replaced waterwheels, but the concept remains the same."

Ms. Howard tells me that structures called tidal barrages capture tidal energy. Tidal barrages are like dams. But instead of harnessing the kinetic energy of a river, they harness the

In the past, humans used tides to power waterwheels. When the tide came in, the moving water rotated the wheel.

kinetic energy of tides. They work best at the mouth of an estuary, where freshwater mixes with seawater. In an estuary, water flows into a narrow channel, which increases its kinetic energy. As the tide rises, the sluice gates of the barrage open. Tidal water flows through the gates and spins the turbines, generating electricity. The sluice gates close, impounding the water, just like a dam. At low tide, the water inside the barrage is higher than outside the barrage. When the sluice gates open, the water rushes out of the barrage back to the ocean, spinning the turbines to generate electricity again.

"Are there any tidal barrages in the United States?" I ask.

"Not yet," Ms. Howard says. "Right now, the only barrages are in a few countries, including France, Canada, and China. It's expensive to build a barrage, and the heavy construction equipment emits greenhouse gases. The barrage also changes

Estuaries are areas where tidal ocean water mixes with freshwater.

water levels and animal life in the tidal area. The water may become cloudy with silt and other particles. And the barrage blocks boats and ships that transport goods. But there are other tidal options!" Ms. Howard adds.

Tidal turbines work very similarly to river turbines. This tidal turbine was installed off the coast of Maine and began supplying power in September 2012.

Another tidal energy option is an offshore lagoon in open water. The offshore lagoon must have a large difference in the height of the water between high and low tide. This works in the same way as the barrage, but it does not affect estuaries at all.

"But the tidal stream option is my favorite," Ms. Howard says. "Tidal energy is very predictable. Two high tides occur every day, approximately 12 hours and 25 minutes apart. Yet, not all locations are suitable for tidal systems. The shape of an area's coastline can change tidal flow and the timing of the tides. The mouth of this river happens to be an ideal location as the river opens into the ocean. Underwater turbines generate electricity using the tides."

Wait. I'm confused. "What's the difference between your tidal stream technology and the river current turbines I visited yesterday?"

Ms. Howard shrugs. "The turbines you saw yesterday use kinetic energy from the river's current, and we use kinetic energy from tides. The size of our turbines and their placement underwater might be different, too, based on the speed and depth of the water. Both kinds of turbines pivot on their posts to catch the flow of the water. But the tidal turbines turn almost 180 degrees to catch the incoming and outgoing tides."

"Do these turbines hurt the fish?" I ask.

Ms. Howard shakes her head. "Our turbines rotate very slowly, similar to river current turbines. At most, our turbines might bump a fish out of the way. Ships and floating debris pass right over."

So we use the same basic technology to harness energy from river currents and tides. But tidal energy isn't the only energy we can get from the ocean.

WHAT CAUSES TIDES?

As the moon rotates around Earth, it exerts some gravitational pull on our planet. As it pulls on the nearest oceans, the moon creates a high tide. On the opposite side of Earth, another high tide occurs. The sun's gravity also tugs at Earth. Its force is weaker than the moon's because it is farther away. About twice a month, during full and new moons, the moon is between the sun and Earth. Together the sun and the moon tug extra hard at Earth's oceans, creating higher-than-normal tides.

CHAPTER 6

Riding the Waves

For my next stop, I'm visiting an experimental wave energy facility in California. According to my research, California, Oregon, Washington, Alaska, and Hawaii have the best wave energy potential in the United States because winds blow from the Pacific Ocean toward the shore. On the East Coast, winds tend to blow away from shore. Scientists calculate the energy in a wave with three measurements: the height of the wave, the length of the wave crest (the top of the wave), and the space between waves. Taller, longer, and faster waves mean more energy.

The surf pounds behind me, and the wind tangles my hair as I introduce myself to Dr. Trevor Roberts. He studies physics, including wave energy. His tanned and lined face reminds me of a veteran surfer.

"Thanks for talking with me today, Dr. Roberts," I say. "I'm hoping you can teach me about wave energy."

Huge waves like this one have a lot of energy. Scientists are working on technology that could harness this energy for human use.

"My favorite topic," he says. "Although wave energy is still experimental, it has the potential to make a terrific contribution to our nation's renewable energy. Particularly here on the West Coast."

"What about Alaska and Hawaii?" I ask.

"Those states have excellent wave potential, but they are far from the other 48 states. Electricity generated by wave technologies would stay in those states to help them reduce their high costs of power. Energy from the West Coast could be used across the country."

Something Dr. Roberts said confuses me. "What are wave technologies?" I ask.

This buoy rises and falls with the ocean's waves. The motion powers a turbine inside the buoy to generate electricity.

"There are five basic technologies to extract energy from waves." He pulls out a chart from his backpack. "One type is point absorbers. These are buoys that float on the water but are attached to the ocean floor. The kinetic energy of the buoys bobbing on the waves becomes electrical energy. Each buoy generates about 500 kilowatts of electricity. Our 200 buoys should be able to generate 100 megawatts of electricity. That's enough to power 24,900 homes."

I look out over the ocean. It's hard to imagine that the waves I'm seeing could produce so much power.

"The second technology is called an overtopping device," Dr. Roberts continues. He points to the second drawing on the chart. "Waves lift water over a barrier and fill a reservoir.

The water in the reservoir drains through a turbine that generates electricity."

"The overtopping device sounds like a floating dam," I say.

"Exactly. You've learned a lot." Dr. Roberts plows ahead and points to the third picture. "Attenuators are devices that look a bit like snakes that ride the waves. Each snake has multiple sections that move with the waves. Attenuators convert the kinetic energy of the sections to electricity."

Dr. Roberts pauses. "Still with me?"

I nod, squinting hard at the pictures.

"The first three devices float on top of the water," Dr. Roberts says, pointing to a fourth picture on his chart. "But oscillating water columns are half in and half out of the water. Above the water, air fills the top part of the device. Incoming waves raise the water level below the surface. This extra water squeezes the air through a turbine, which spins to generate

WIND FORECASTING

You've probably heard of weather forecasting, but what about wind forecasting? A program at the Hawaii National Marine Renewable Energy Center develops models of wind patterns. Wind makes waves. If scientists can determine when wave-making winds will blow, they will be able to determine the best locations for wave-operated power systems. To understand the winds around the Hawaiian Islands, scientists are examining past wind patterns and predictions of future wind patterns to create a huge bank of information for renewable energy experts.

electricity. As the strength of the wave passes, the water level in the device decreases. Less water decreases the air pressure. The air is pulled through the turbine for more electricity."

The last picture on Dr. Roberts' chart looks like an upside-down pendulum. "Let me guess what this one does," I say.

He hands me the chart. "Give it a whirl!"

"The upside-down pendulum moves with the back-and-forth motion of the waves."

Dr. Roberts nods. "We call that surge."

"Okay, the waves surge and force the lever back and forth. The lever is connected to a motor. The motor turns kinetic energy into mechanical energy. Then a generator turns the mechanical energy into electricity."

Dr. Roberts smiles. "How would you like to work for us in a few years?"

Waves are a source of kinetic energy. This energy can be captured in many different ways.

CHAPTER 7

Shallow/Warm, Deep/Cold

Today I'm visiting a renewable energy center in Hawaii. I want to ask the center's manager, Scott Meyers, some questions about a brand-new type of ocean energy.

"Welcome to our pilot plant," Mr. Meyers says. The plant is right next to the beach, and he is wearing a colorful Hawaiian shirt and flip-flops. "We're testing an exciting new type of hydropower at our facility—OTEC."

"Does OTEC stand for something special?" I ask.

"**O**cean **t**hermal **e**nergy **c**onversion." Mr. Meyers grins at my blank look. "Tides and waves are not the only sources of energy in the ocean. OTEC uses the temperature difference between warm water at the ocean's surface and cold water 3,000 feet, or 1,000 meters, below. Our experimental plant has the ability to produce steady energy without waiting for tides to ebb and flow."

We can harness energy from more than just the ocean's waves.

I look out to sea. I remember 96.5 percent of the earth's water is in its oceans. "Why isn't OTEC used all over the world?"

"Excellent question," Mr. Meyers says. "For OTEC to work, the surface water needs to be at least 36 degrees Fahrenheit, or 20 degrees Celsius, warmer than the deep water. While our oceans are full of areas that could work for OTEC, most of them are too far from shore. The distance makes the conversion to electricity expensive and difficult. Florida, Hawaii, and the Pacific Islands have the best chance of success. Seasons make a difference, too. Surface water is colder in the winter, so potential OTEC areas are even fewer."

"I get it now, Mr. Meyers, but how do you make electricity from the cold and warm ocean water?"

Mr. Meyers kicks four piles of sand together to illustrate his explanation. He points to the first pile of sand. "Here we have a container for warm surface seawater. First, we pump the seawater in. Then we send it to a second container called a flash evaporator that reduces the pressure." He gestures to the second pile of sand.

"As the pressure drops, the water vaporizes to steam. The volume of the steam is now greater than the seawater. It needs to move out of its small container." He points to the third pile of sand. "The steam blasts past the turbine that spins to generate electricity, and into a fourth container. Here, cold seawater pumped from more than 3,000 feet, or 1,000 meters, below surrounds the last container and condenses the steam back to salt-free water."

"Salt-free water?" I pause to think about what Mr. Meyers is telling me. "So you're saying that ocean thermal energy conversion not only generates electricity, but it also creates drinking water?"

HYDROTHERMAL VENTS

The earth's crust is made up of giant plates. Since oceans cover much of Earth's surface, many of these plates meet underwater. As the plates shift, small cracks form. Seawater trickles through the cracks to the earth's core of hot magma. The seawater becomes super-heated to 700 degrees Fahrenheit (370°C) and returns through hydrothermal vents 7,500 feet (2,300 m) deep. The extreme pressure at these depths prevents the water from boiling. If we can channel the energy of hydrothermal vents to the surface, they might become the next source of renewable water power.

Mr. Meyers claps me on the back. "That's exactly what OTEC does!"

Wow! I knew water was important to humans' survival, but I had no idea we could use water power for so much. Like all energy sources, water power has its pros and cons. But it is clean and renewable. It is already an important source of energy for people across the United States and around the world. I'm excited to see what the future holds for water power.

One day we may be able to use hot water spewed out by hydrothermal vents deep in the ocean to generate electricity.

CHAPTER 8

Your Turn

You've had a chance to follow Zach as he conducted his research. Now it's time to think about what you've learned. You have studied pros and cons of dams and tidal barrages. These structures can produce a lot of power, but they can also change the environment in which they are built. Free-flowing systems such as river currents and tidal turbines don't have as many negative affects on the environment, but they don't produce nearly enough power on their own to meet our energy needs. You have also learned about experimental ocean energy technologies, such as waves and ocean thermal energy conversions. While these don't play a large role in our water power use right now, they may be an important part of our energy future. Scientists and engineers keep exploring new and better ways to make use of water power. Maybe you'll make the next big discovery!

YOU DECIDE

1. Do you think the positive aspects of water power outweigh the negative aspects? Why or why not?

2. Which water power technology discussed in Zach's journal do you think is most promising? Why?

3. What can you do to cut down on your energy use? Think about technology and behavior changes.

4. If you had enough money to invest in one type of water power research, which project would you choose? Why?

5. How big of a role do you think water power will play in our future energy production?

Water is needed for life on Earth. It has played an important role in our energy production for thousands of years. Soon, it may have an even greater role.

GLOSSARY

attenuators: Ocean energy devices that capture wave energy to convert it to electricity.

barrage: A low dam built across an estuary to capture the energy produced by tides.

estuary: A body of water formed by freshwater rivers and streams emptying into the salt water of the ocean.

fossil fuels: Energy sources that developed from the remains of animals and plants.

generator: A machine that converts mechanical energy into electricity.

gravity: The force that attracts a body toward the center of the earth.

hydroelectric: Generating electricity from flowing water that drives a turbine.

kilowatts: One thousand watts of electrical power.

kinetic energy: The energy a substance in motion contains.

megawatts: One million watts of electrical power, or 1,000 kilowatts.

non-renewable: When something cannot be replaced by natural environmental cycles as quickly as we are using it.

oscillating: Moving or swinging back and forth in a regular rhythm.

penstock: Large tubes that lead water from a reservoir above a dam to the turbines that spin to generate electricity.

renewable: When something can be replaced by natural cycles in nature or the environment.

reservoir: A place where something is collected, especially a liquid.

sluice gates: A sliding gate that controls the flow of water.

tributaries: Streams or branches of a river that feed into a larger river.

turbine: A machine that spins to power a machine that generates electricity.

voltage: The measurement of the energy contained within an electric circuit at a given point.

EXPLORE FURTHER

Make a Turbine

You will need an empty cardboard half-gallon milk carton. Use a nail to punch one hole in the exact middle of the top of the carton. Put a piece of string through the hole so you can hang the carton. Punch two additional holes in the bottom right corner and the bottom left corner. Put heavy tape over these two holes. Hang the carton from a tree outside. Fill the carton with water. Rip the tape off one corner. What happens? Rip the tape off the other corner. What happens? Explain what is happening with the concepts you learned in this book.

Where Does the Water Cycle Begin?

Show your friends, teachers, parents, and coaches the picture of the water cycle in Chapter Three of this book. Ask each person the following questions:

- Where does the water cycle begin?

- What is the best source of hydropower?

- Make one chart for each question summarizing the responses. Graph your results.

Do You Conserve or Waste Energy at Home?

Draw a map of your home. Visit each room to find the items that use energy, such as lights, the thermostat to regulate heat or air conditioning, televisions, computers, appliances, etc. Indicate whether these items were on or off when you visited each room. Ask your parents if your home uses renewable forms of energy, such as solar power. Can you think of ways to conserve energy in your home?

What can you do in your home to help conserve energy?

SELECTED BIBLIOGRAPHY

Doeden, Matt. *Green Energy: Crucial Gains or Economic Strains*. Minneapolis: Twenty-First Century Books, 2010. Print.

NOAA. "Tides and Water Levels." *NOAA Ocean Service Education*. National Oceanic and Atmospheric Administration, March 25, 2008. Web. Accessed September 27, 2012.

Quick, Darren. "Autonomous Wave Energy PowerBuoy Device Commences Sea Trial." *Gizmag*, August 23, 2011. Web. Accessed June 1, 2012.

U.S. Department of the Interior, Bureau of Reclamation. "Ecosystem Restoration." *Reclamation: Managing Water in the West*. U.S. Department of the Interior, n.d. Web. Accessed June 4, 2012.

FURTHER INFORMATION

Books

Gardner, Robert. *Energy: Green Science Projects about Solar, Wind, and Water Power*. Berkeley Heights, NJ: Enslow, 2011.

Goodman, Polly. *Understanding Water Power*. Pleasantville, NY: Gareth Stevens, 2011.

Sechrist, Darren. *Powerful Planet: Can Earth's Renewable Energy Save Our Future?* Pleasantville, NY: Gareth Stevens, 2010.

Spilsbury, Richard. *Water, Wave, and Tidal Power*. New York: PowerKids Press, 2012.

Websites

http://www.bpa.gov/PublicInvolvement/ CommunityEducation/ValueoftheRiver/Pages/default.aspx
The Bonneville Power Administration teaches you about a dam on the Columbia River.

http://energyquest.ca.gov/index.html
The Energy Quest website contains tons of great information on energy, including a story, movies, and more.

http://www.eia.gov/kids/energy.cfm?page=hydropower_ home-basics
This website just for students discusses many of the sources of water power described in this book. Be sure to look at the pictures!

INDEX